G. M. Hopkins

# Atlas of the county of Montgomery and the state of Pennsylvania

G. M. Hopkins

**Atlas of the county of Montgomery and the state of Pennsylvania**

ISBN/EAN: 9783742835628

Manufactured in Europe, USA, Canada, Australia, Japa

Cover: Foto ©Klaus-Uwe Gerhardt /pixelio.de

Manufactured and distributed by brebook publishing software
(www.brebook.com)

G. M. Hopkins

# Atlas of the county of Montgomery and the state of Pennsylvania

# TABLE OF CONTENTS

## TABLE OF DISTANCES
OF
MONTGOMERY COUNTY.

Showing the distances between Villages and Boroughs
in miles and tenths of miles.

MEASURED ON THE NEAREST PUBLIC ROADS.

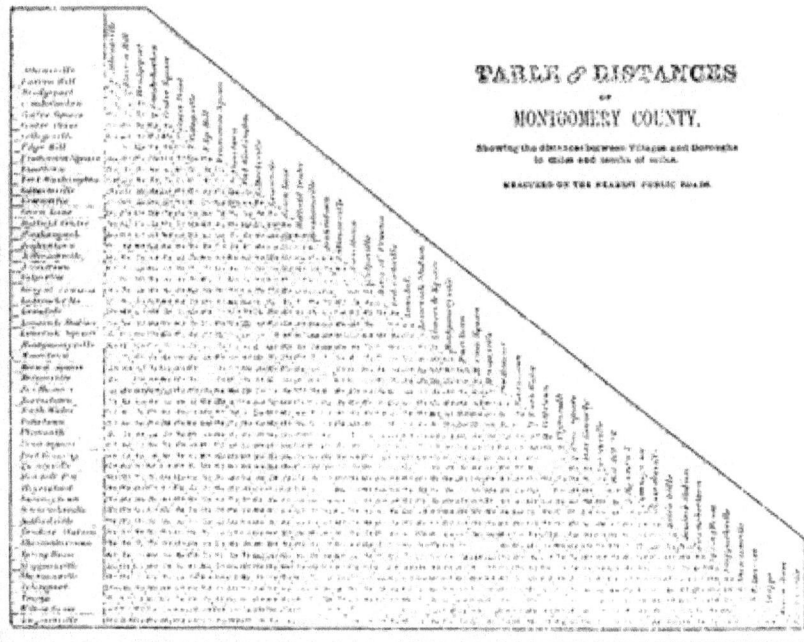

# TABLE of Air Line Distances

IN MILES AND TENTHS OF MILE,

Between the different County Seats thro'out the
State of Pennsylvania.

OUTLINE MAP
OF
MONTGOMERY CO.
by
G. M. HOPKINS, C.E.

Scale 2½ miles per inch.

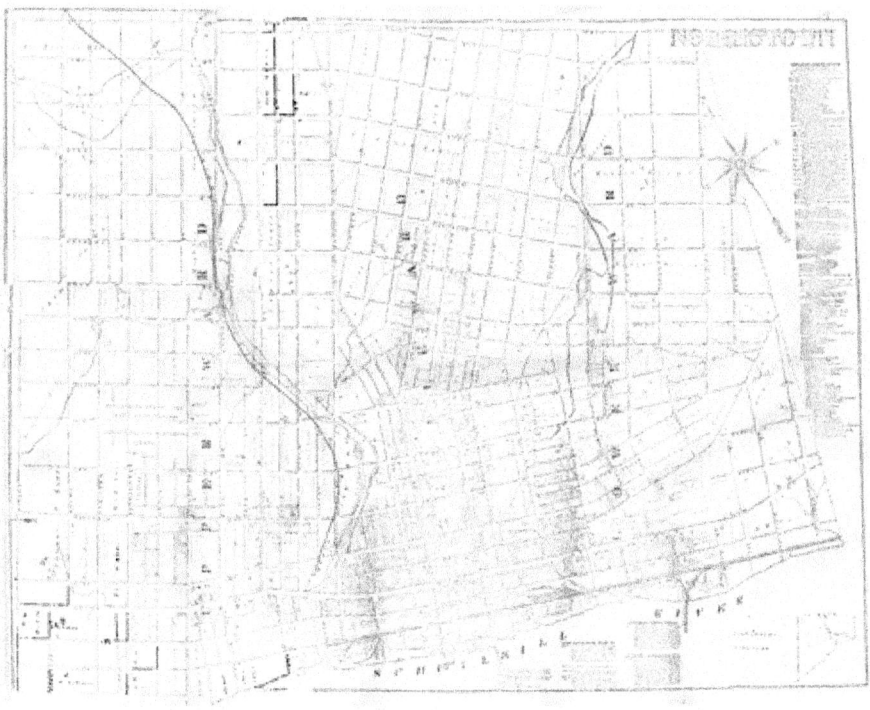

# MORELAND.

Scale 2 inches per Mile

WEST CONSHOHOCKEN.

Scale 500 ft. per Inch.

SCHUYLKILL RIVER

# HORSHAM.

Scale: 2 Inches to the Mile.

# MARBLE HALL.

# BARREN HILL.

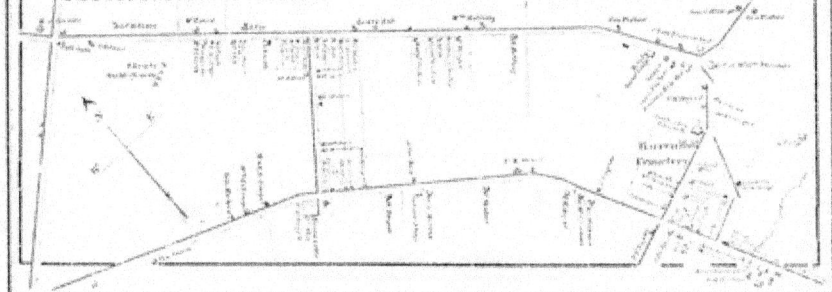

# GWYNED

Scale 2 inches per Mile

# BOROUGH
OF
NORTH WALES.

### North Wales Business Directory.

### GWYNEDD BUSINESS DIRECTORY.

## Plan of
## FLOURTOWN.

Scale 200 feet per inch

## Plan of
## LANSDALE.

Scale 200 feet per inch

BUCKS COUNTY

DIST No 3

MONTGOMERYVILLE

MONTGOMERY SQUARE

DIST No 2

# MONTGOMERY.

Scale 2 Inches per Mile.

Montgomery Township Business Directory

LINE LEXINGTON

HOUGHTOWN

HATFIELD SQUARE

DIST No 2

HATFIELD CENTRE

DIST No 3

DIST No 4

LANSDALE

# HATFIELD.

Scale 2 Inches per Mile.

## HATFIELD BUSINESS DIRECTORY.

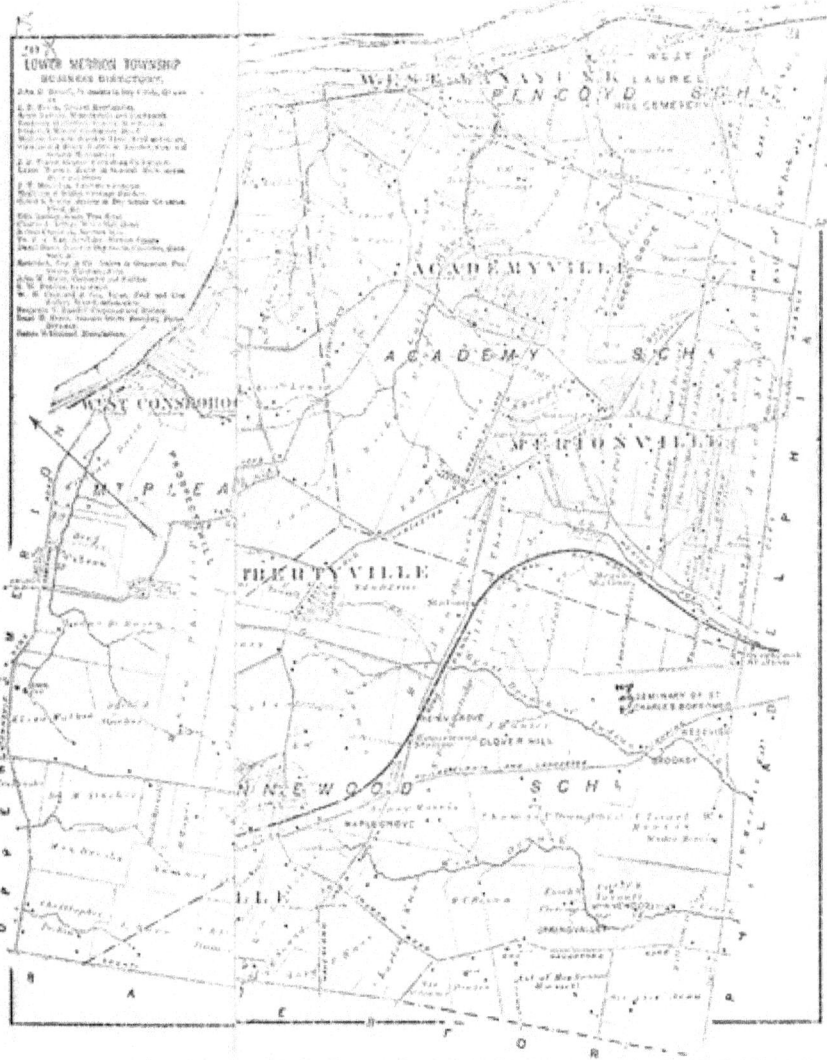

LOWER MERION TOWNSHIP
BUSINESS DIRECTORY.

WEST
WYNNEWOOD
PENCOYD
LAUREL
HILL CEMETERY

ACADEMYVILLE

ACADEMY SCH.

MERIONVILLE

WEST CONSHOHOCKEN

MAPLE

LIBERTYVILLE

SEMINARY OF ST.
CHARLES BORROMEO

CLOVER HILL

WYNNEWOOD SCH.

MAPLEGROVE

SPRINGVILLE

LOWER MERION.

# UPPER MERION.

Scale—2 Inches per Mile.

**Upper Merion Business Directory.**

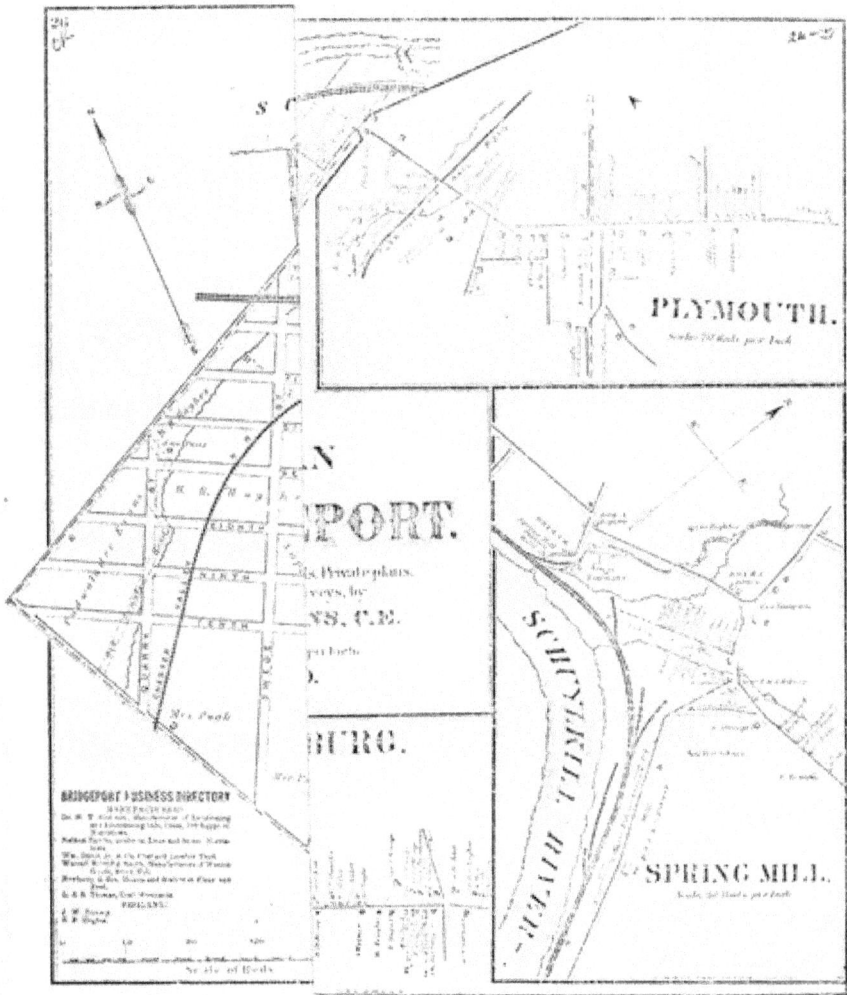

PLYMOUTH.

Scale 20 Rods per Inch

...PORT.

...NS, C.E.

...BURG.

SPRING MILL.

Scale 20 Rods per Inch

SCHUYLKILL RIVER

BRIDGEPORT BUSINESS DIRECTORY

Scale of Rods

PLAN
of
BRIDGEPORT.

SCHUYLKILL RIVER

PLYMOUTH

PENNSBURG.

GREENVILLE.

SPRING MILL.

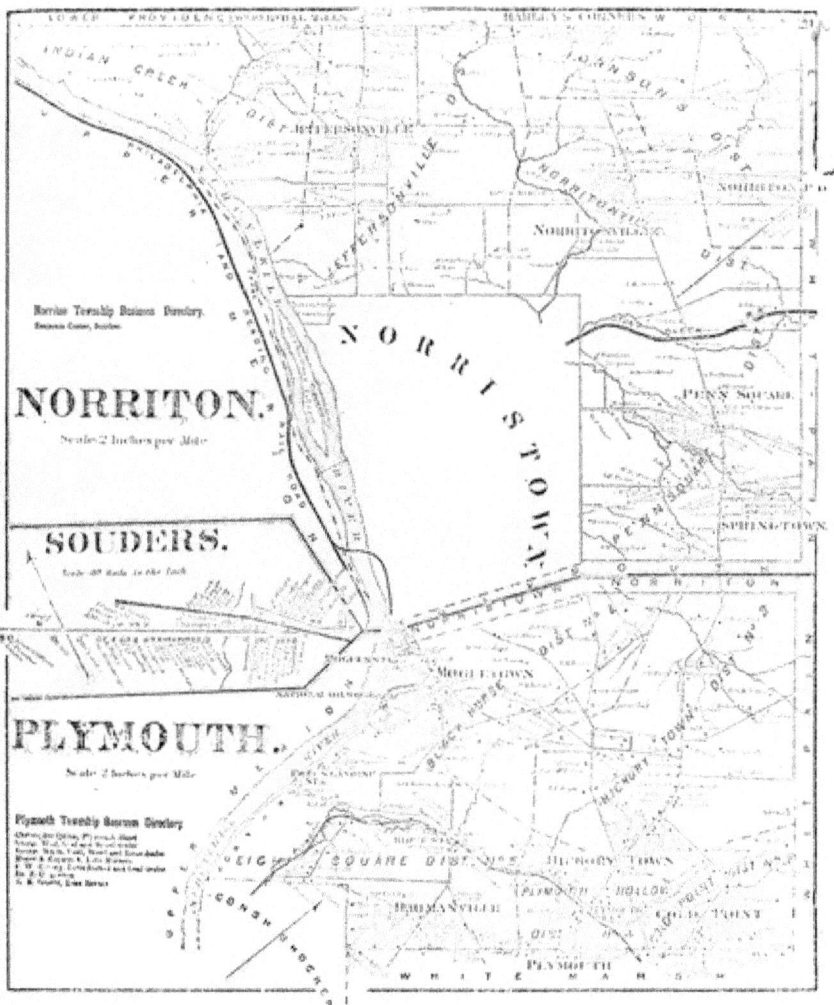

# NORRITON.

Scale 2 Inches per Mile

Norriton Township Business Directory.

SOUDERS.

Scale 40 Rods to the Inch

# PLYMOUTH.

Scale 2 Inches per Mile

Plymouth Township Business Directory.

NORRISTOWN

NORRITON

PENN SQUARE

SPRINGTOWN

JEFFERSONVILLE

NORRITONVILLE

NORRISVILLE

INDIAN CREEK

MOGEETOWN

HICKORY TOWN

PLYMOUTH HOLLOW

COLD POINT

HARMANVILLE

CONSHOHOCKEN

WHITEMARSH

PLYMOUTH

SQUARE DIST. NO. 2

BLACK HORSE

# WHIT PAIN.

Scale: 2 inches per Mile.

C. M. HOPKINS.

Whitpain Business Directory.

# FRANCONIA.

Scale: 2 inches per Mile.

FRANCONIA TOWNSHIP
BUSINESS DIRECTORY.

# UPPER PROVIDENCE.

Scale 2 inches per Mile.

## PORT PROVIDENCE.

## QUINCYVILLE.

35

PERKIOMEN — WORCESTER

PROVIDENCE SQUARE

EVANSBURG DIST

TRAPPE DIST

CHERRY TREE DIST

PERKIOMEN BRIDGE

MOUNT KIRK

MAPLE TREE DIST

THE LEVEL DIST

THE HOLLOW DIST

THE BEACH DIST

SHANNONVILLE

# LOWER PROVIDENCE.

Scale 2 Inches per Mile

WETHERILLS CORNER

JACK DIST

UPPER MERION

## UPPER PROVIDENCE DIRECTORY.

## Lower Providence Business Directory

# LIMERICK.

Scale 2 Inches per Mile

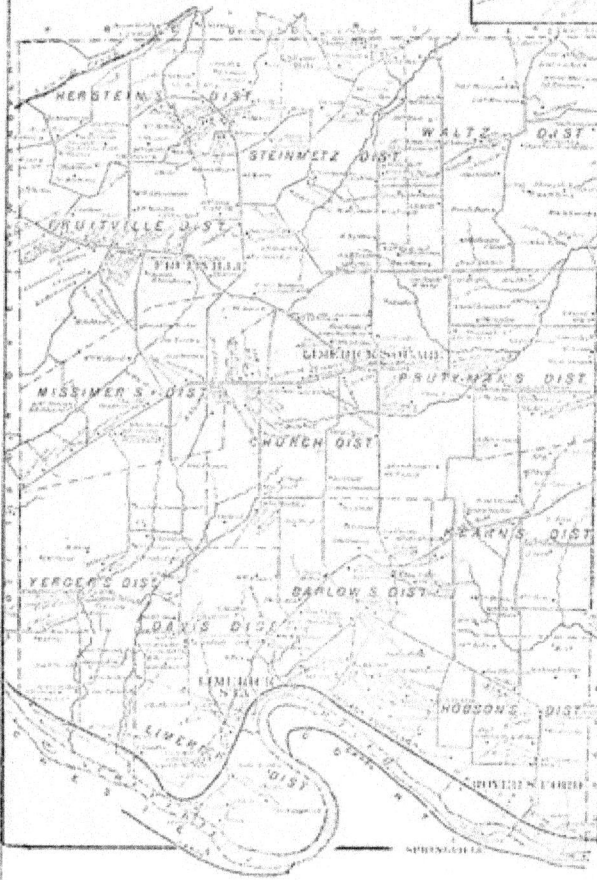

LIMERICK SQUARE

HERGSTEIN'S DIST

STEINMETZ DIST

WALTZ DIST

FRUITVILLE DIST

LIMERICK SQUARE

MISSIMER'S DIST

CHURCH DIST

PRUTZMAN'S DIST

ROYERSFORD

EARN'S DIST

YERGER'S DIST

BARLOW'S DIST

DAVIS DIST

LIMERICK

LIMERICK

HOBSON'S DIST

LIMERICK STATION.

ROYERSFORD

SPRINGVILLE

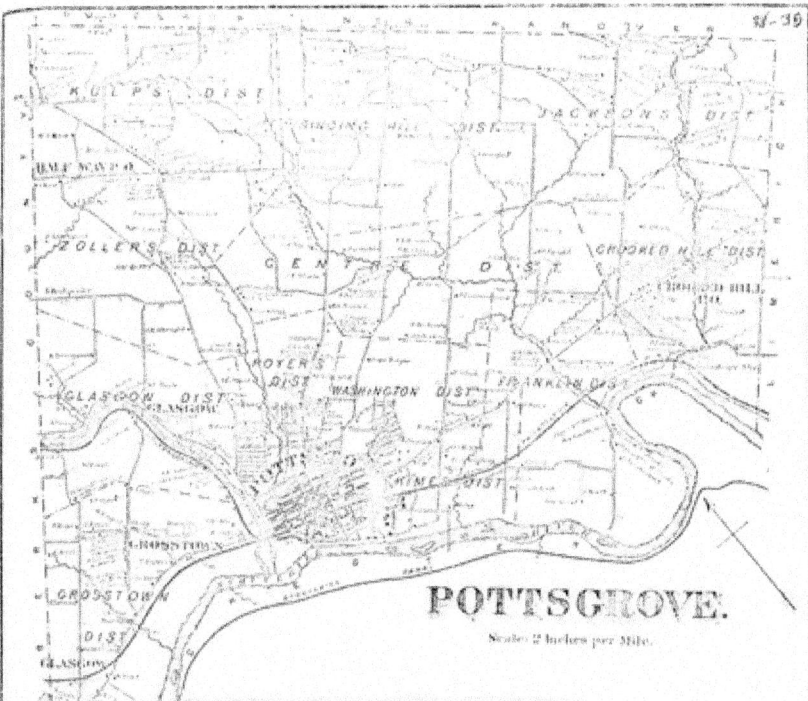

# POTTSGROVE.

Scale 2 Inches per Mile.

LIMERICK DIRECTORY.

POTTSGROVE DIRECTORY.

**PLAN OF**
THE BOROUGH OF
**POTTSTOWN**
MONTGOMERY CO. PA.

G. M. HOPKINS, C. E.

SCHUYLKILL RIVER

DOUGLASS.

Scale: 2 inches per mile.

GILBERTSVILLE.

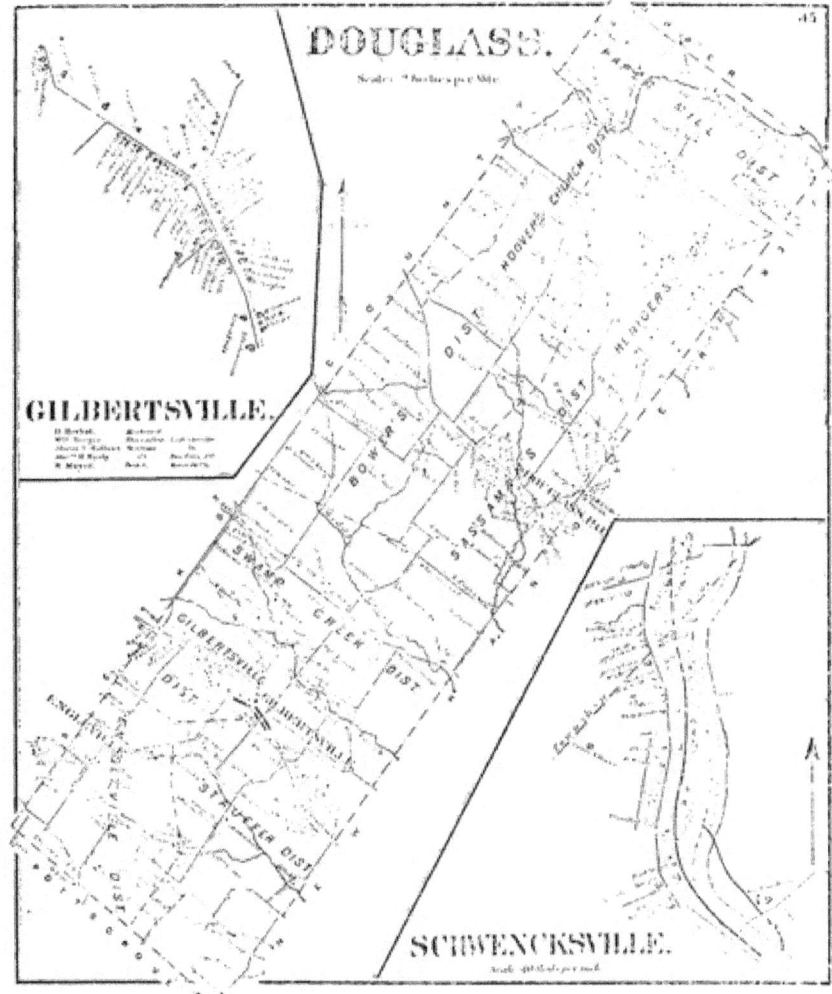

SCHWENCKSVILLE.

## KULPSVILLE.

DENNIS DIST.

UNION SQUARE

KRUPPS DIST. No 2

KULPSVILLE

DIST No 5    DIST No 6

KREIBELS' DIST.

LOCUST CORNER

### Business Directory.

UPPER SALFORD
DIRECTORY.

**TOWAMENCIN.**

Scale 2 Inches per Mile

## UPPER
## SALFORD

# PLAN OF

## THE BOROUGH OF

# CONSHOHOCKEN.

## MONTGOMERY CO. PA.

From Official Records, Private Plans and Actual Surveys by
### H.W. HOPKINS, C. E.

Drawn by
### G. M. HOPKINS, C. E,
320 WALNUT ST. PHILADELPHIA.

Scale: 300 Ft. to an Inch.

CONSHOHOCKEN BUSINESS DIRECTORY
John Wood & Bros
Charles A. Harrett,
John P. Jones.

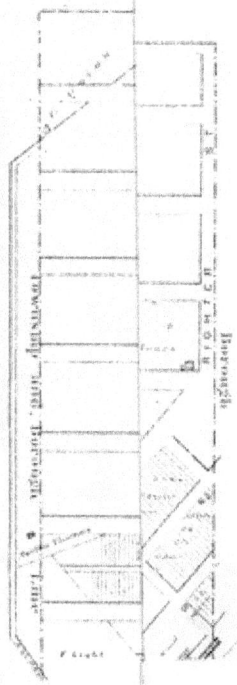

PLAN OF
THE BOROUGH OF
CONSHOHOCKEN.
MONTGOMERY CO. PA.

From Office and Private Plans and Actual Surveys by
B. W. HOPKINS, C. E.

Drawn by
G. M. HOPKINS, C. E.
320 WALNUT ST. PHILADELPHIA.

SCHUYLKILL RIVER

## ALLEGVILLE.

## UPPER DUBLIN.

Scale 2 Inches per Mile

### DANIEL, &c. DIRECTORY.

### Upper Dublin Business Directory.

PENNVILLE.

SCHWENKSVILLE P.O. N 4

AMBLERTOWN

DRESHERTOWN

FITZWATERTOWN

PINETOWN

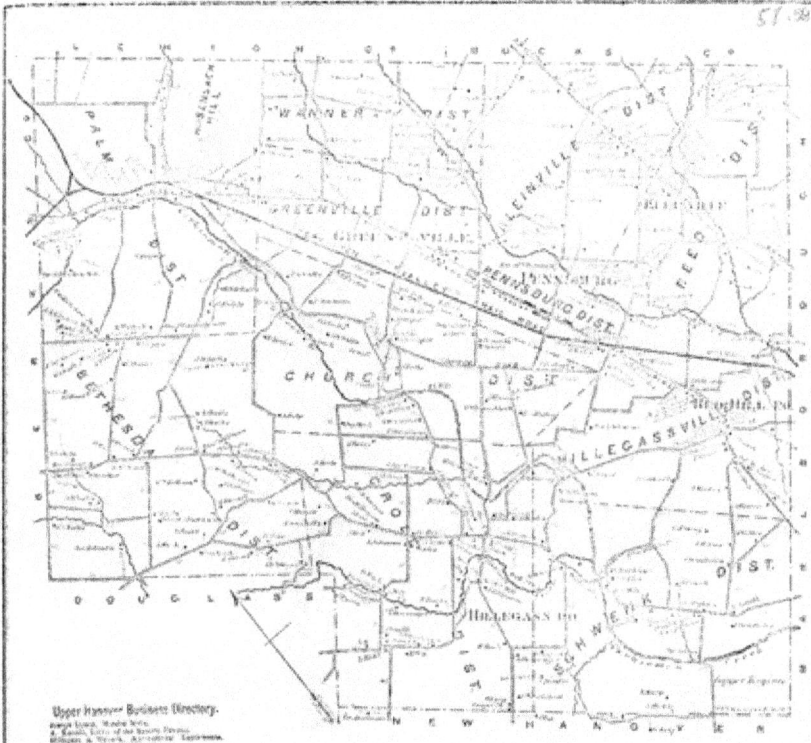

UPPER HANOVER.

Scale 2 Inches per Mile.

MOITE TOWN

A B I N G T O N

JENKINTOWN

WELDON

EDGEHILL

WELDED HEIGHTS

SHOEMAKERTOWN

C H E L T E N H A M

CHELTENHAM

CITY OF PHILADELPHIA

# ABINGTON
## AND
# CHELTENHAM.

Scale 2 Inches per Mile.

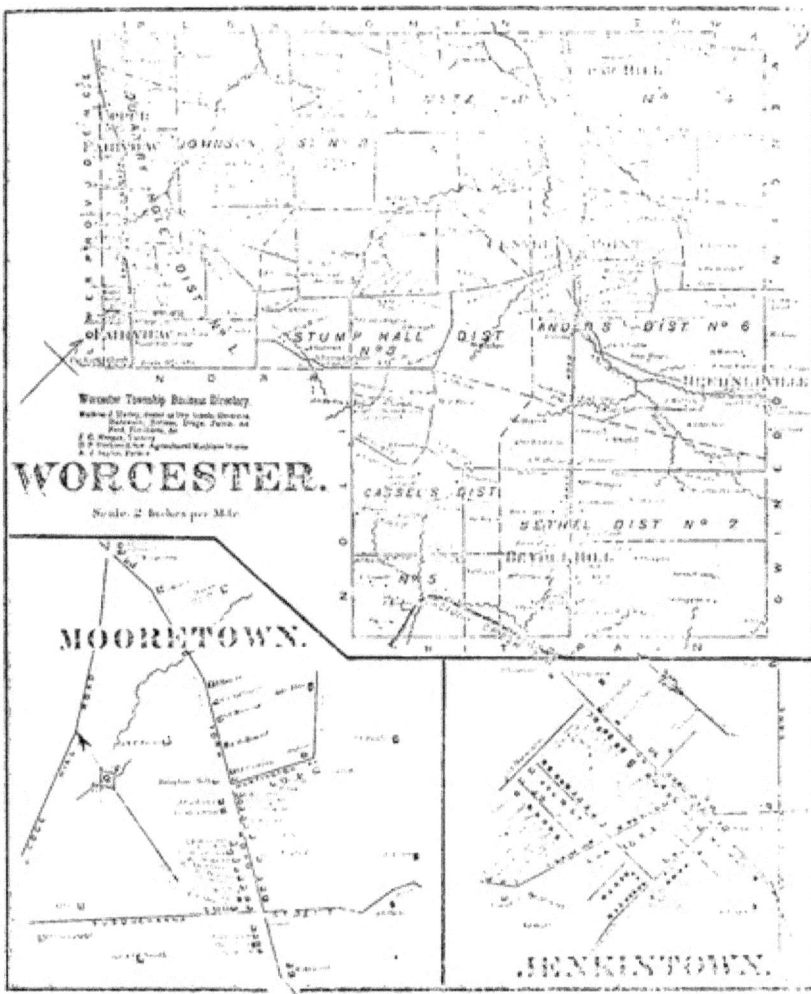

# WORCESTER.

Scale 2 inches per Mile

Worcester Township Business Directory.

WORTH PROVIDENCE

FAIRVIEW  JOHNSON'S  DIST No 1

DIST No 1

STUMP HALL DIST No 3

SANDERS DIST No 6

HIGHENLONVILLE

CASSEL'S DIST

BETHEL DIST No 7

DEVOL HILL

No 5

# MOORETOWN.

# JENKINTOWN.

# HARLEYSVILLE

## FREDERICK DIRECTORY.

# FREDERICK.

Scale of Inches per Mile.

GOSHENVILLE P.O.

JONES DIST.

GRIMLY'S DIST.

ZIEGLERSVILLE  OBERNVILLE

DIST.

DIST.

LEIDIG'S DIST.

STONE HILL DIST.

SPRING MILL

MARBLE HALL

COLD POINT

UPPER

LOWER

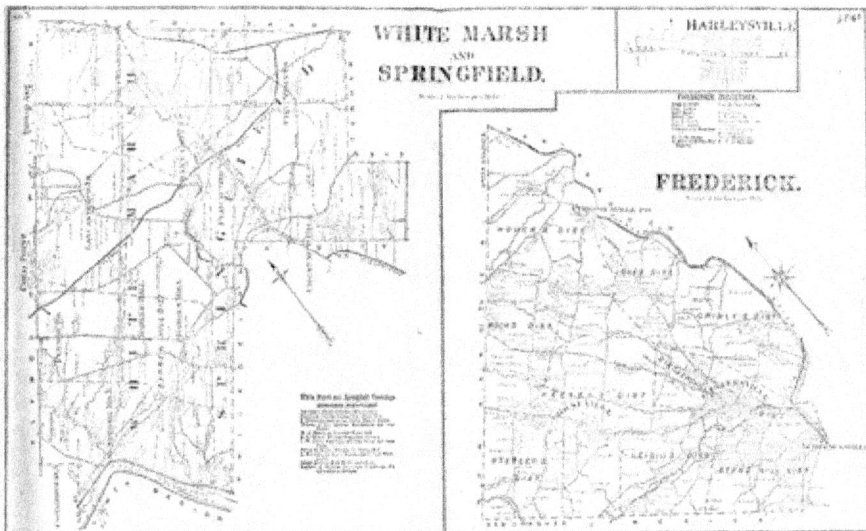

WHITE MARSH
AND
SPRINGFIELD.

HARLEYSVILLE

FREDERICK.

WEST
LAUREL HILL
CEMETERY,
on
BELMONT AVENUE
beyond West Park.
Office of the Company:
No 324 WALNUT ST. PHILAD.ª
Laid out by G. B. DOWNS, C.E., 3209 Race St.
PHILADELPHIA, Pª
1871.
Scale.

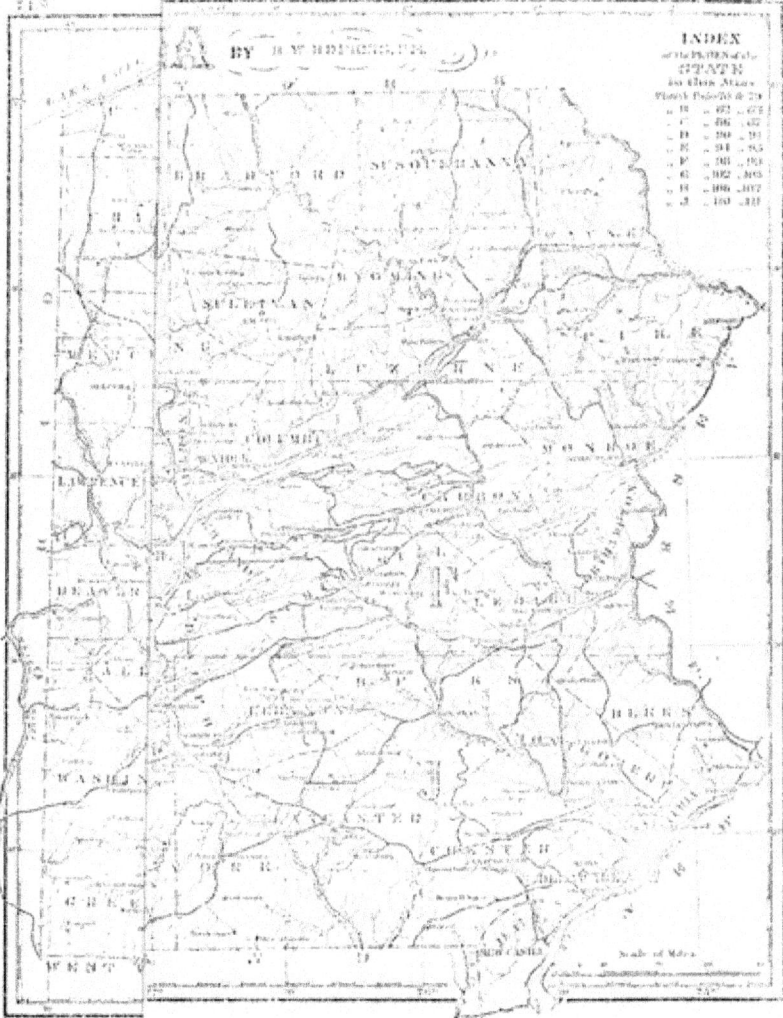

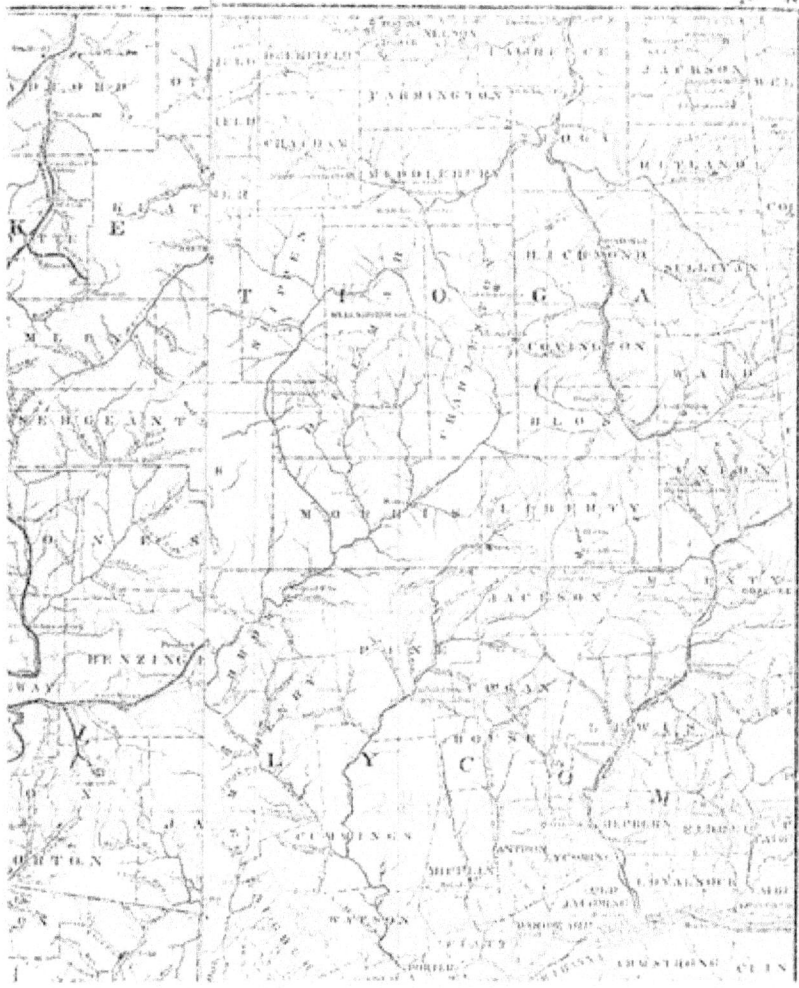

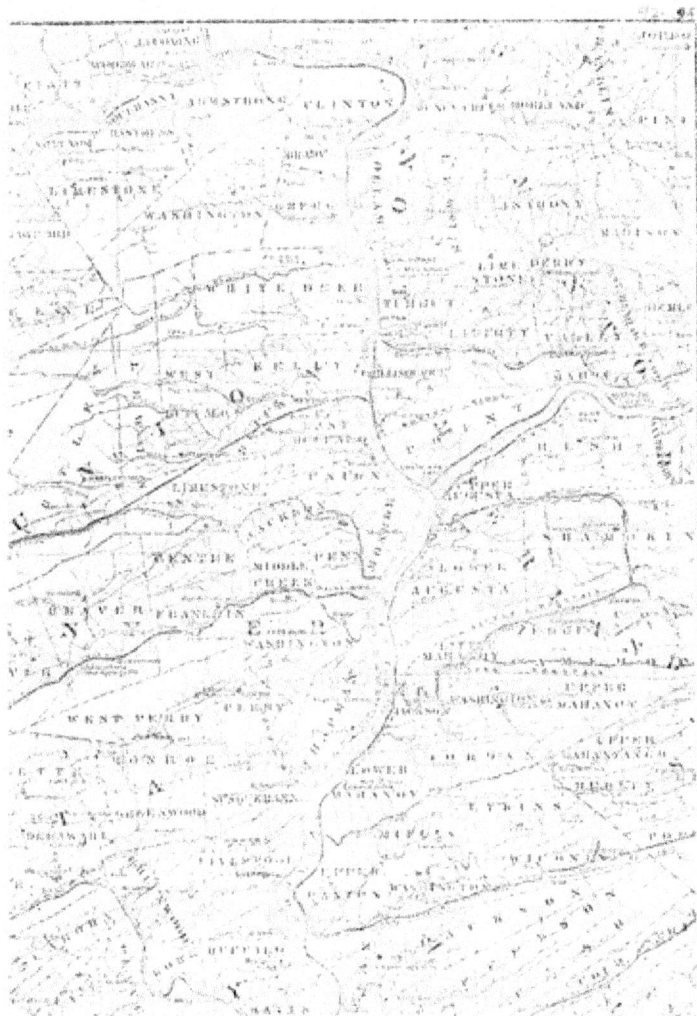

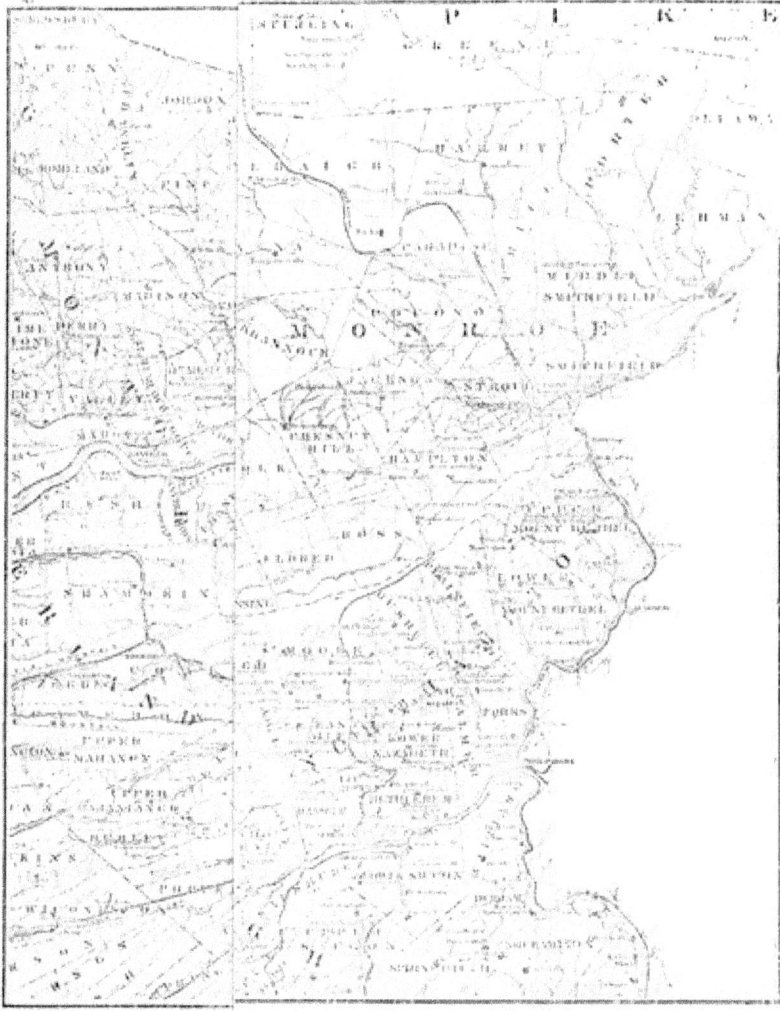

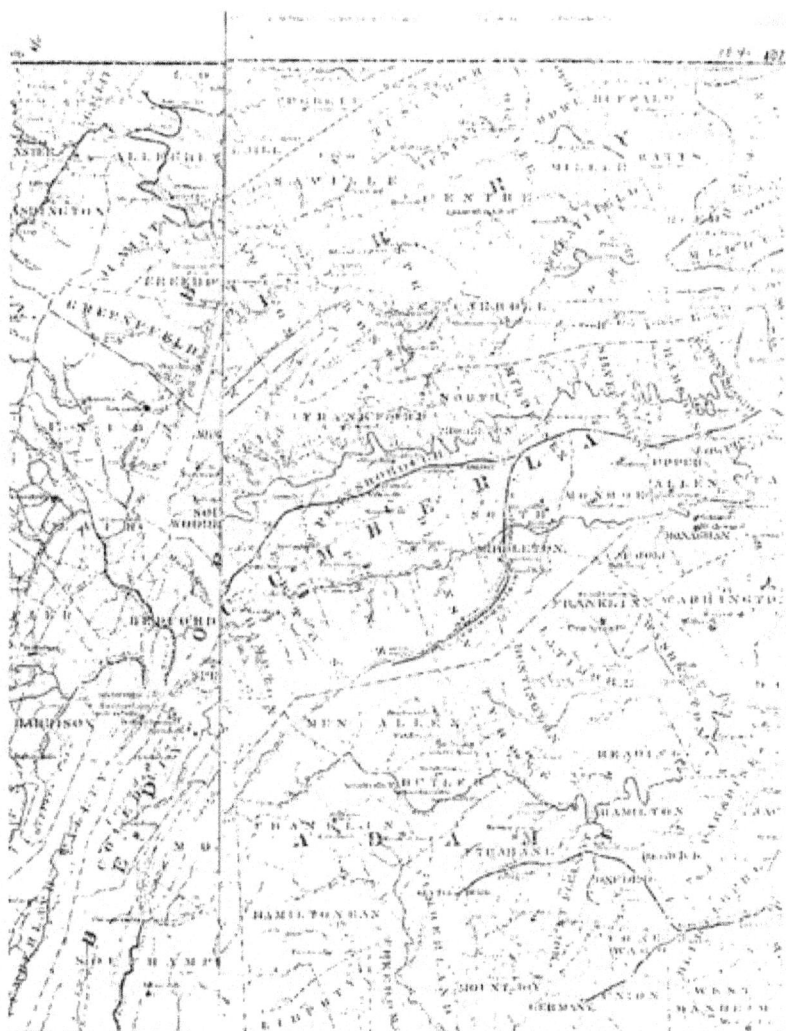

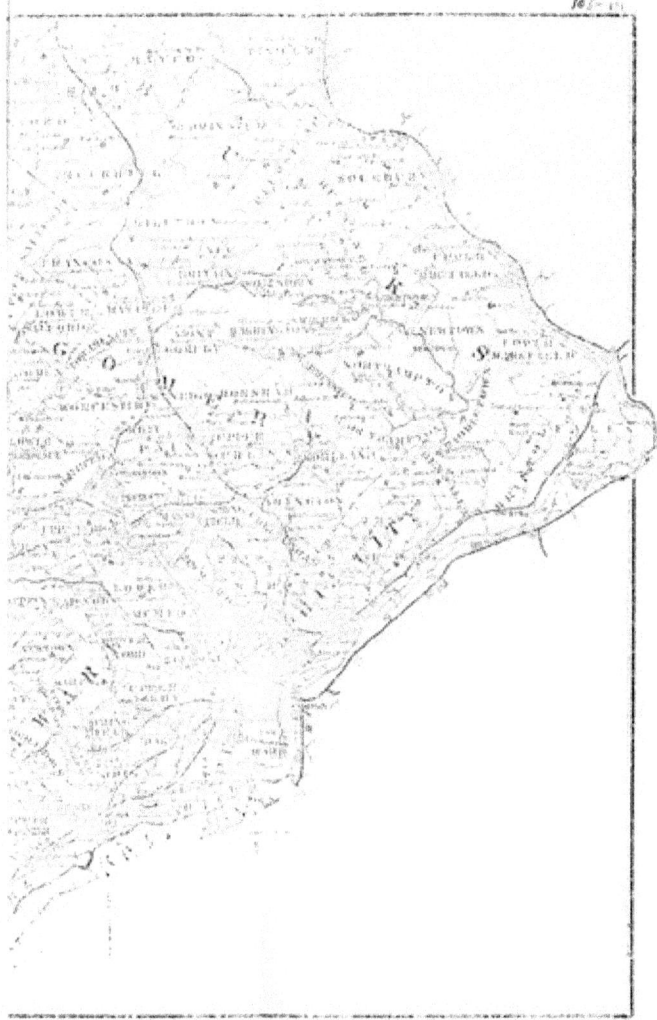